Curious Kids

Create Limitless Learning

Know, Want to Know, Learned (KWL)

By

DeLisa R Jones, M. Ed.

Dedication

This book is dedicated to my grandchildren (Monika, Desaray, Courtney, Kervin (RIP), and Madisyn) and to all the scholars whom I have been blessed to stand before as an educator, to my beautiful foster children whom God has entrusted in my care. Finally, I dedicate this writing to my wonderful adult children, Angel & Murshid. Thanks, to each of you for putting up with me and loving me. To God be the Glory.

Words of Praise for Curious Kids

“Curious Kids” is a book that aims to inspire children to read by using various topics and interests as a means of engagement. The book incorporates questions that children often ask their teachers in the classroom, such as who’s the best, the tallest, or the fastest. By invoking children’s curiosity, the book encourages lifelong learning habits.

The author, Ms. Jones, uses fun facts as the primary tool to capture the attention of young readers. She then takes them on a journey through a series of short fact-based readings about ethnic inventors, followed by a profile of the “Fastest Man Alive.” Other sections of the book explore figurative language, reading strategies, and poetry, including some of Ms. Jones’s own poems. The book ends with a “Did You Know?” collection of intuitive questions and answers, and a “Do You Know?” section that encourages personal input from the reader.

One of the most important techniques used in “Curious Kids” is the KWL chart, which stands for what students already know, what they want to know, and what they ultimately learn. This graphical organizer is designed to help students learn by constructing their own understanding of a topic or idea. This constructivist teaching method is highly effective and can be used with any text at any level.

In a post-COVID educational environment, where young readers may have fallen behind, “Curious Kids” can be a valuable resource for motivating young learners and facilitating their growth.

--DEBORAH D. STOKES

Secondary Educator for 38 years

“Curious Kids” will be an invaluable tool for its readers. It offers a unique blend of educational practices and personal contributions, such as Jones’s poems, that come straight from the heart. The book is both intelligent and heartfelt, a testament to the passion that went into its creation. I particularly appreciate the “KWL” approach, which encourages readers to approach new information thoughtfully and systematically. Jones is clearly passionate about knowledge, children, and God, and this passion is infectious. I have no doubt that this book will inspire and inform young readers, who are, after all, the future of our world.

-- GABRIELLE WOMAC, BS, CNIM

Table of Contents

Fun Facts

Who Is Considered the Greatest Athlete of All Time (Goat)?

Bo Jackson is the greatest athlete of all time (GOAT). He was an All-Star in both Major League Baseball and the National Football League. He is the only person to be named an All-Star in two different professional sports in the United States.

Who is the Tallest Man on Earth Right Now?

Sultan Kösen (born 10 December 1982) is a Turkish Farmer, who holds the Guinness World Record for tallest living male at 251 cm (8 ft 2.82in). He is of Kurdish ethnicity and listed as the tallest individual. Kösen's growth resulted from the conditions of gigantism and acromegaly, caused by a tumor affecting his pituitary gland.

What is the Most Famous Artificial Waterway?

The Panama Canal is one of the most important waterways in the world. Without it, ships sailing from the east coast in the United States would be forced to make the two-week-long journey around Cape Horn in southern Chile. Extending for about 50 miles, the canal connects the Pacific Ocean to the Atlantic Ocean and divides North America and South America.

Can Dogs Understand Words?

According to the American Psychological Association, the average dog can understand 165 words, and "super dogs"—those in the top 20% of canine intellect—can understand around 250 words. Dog intelligence can be divided into three main types: instinctive (what the dog is bred to do), adaptive (what a dog learns from its environment), and working/obedience (what a dog is trained to do). Research into the levels of working/obedience intelligence in various dog breeds shows that border collies displayed the highest levels, followed by poodles, German shepherds, and golden retrievers. With the ability to also understand simple math (for instance, 1+1=2), these "super dogs" have an estimated cognitive ability of 2-2.5-year-old humans.

Can Cats Understand Human Words?

Cats lack the cognitive skills to interpret human language, but they recognize it when you talk to them. To put it another way, cats comprehend human language the same way we understand meowing. It is like how you interpret your cat's language by "reading" how they arch their back or swish their tail. Cats may only understand 25-35 words, but they can make about 100 different vocalizations. Perhaps, cats are desperately trying to make first contact with us, as they only make these sounds around their human friends, not around other cats.

What are the Tallest Trees in the World?

The world's largest trees are the Hyperion, which is a coastal redwood (Sequoia Sempervirens). It is located somewhere in the heart of Redwood National Park in California.

Where Does the Sky End and Space Begin? The Kàmàn Line!

Our atmosphere doesn't end in one spot; it thins out the higher you get, until there is almost no air left at all. For simplicity, scientists say that the atmosphere ends at the Kάmάn line, 100 km (62 miles) above sea level.

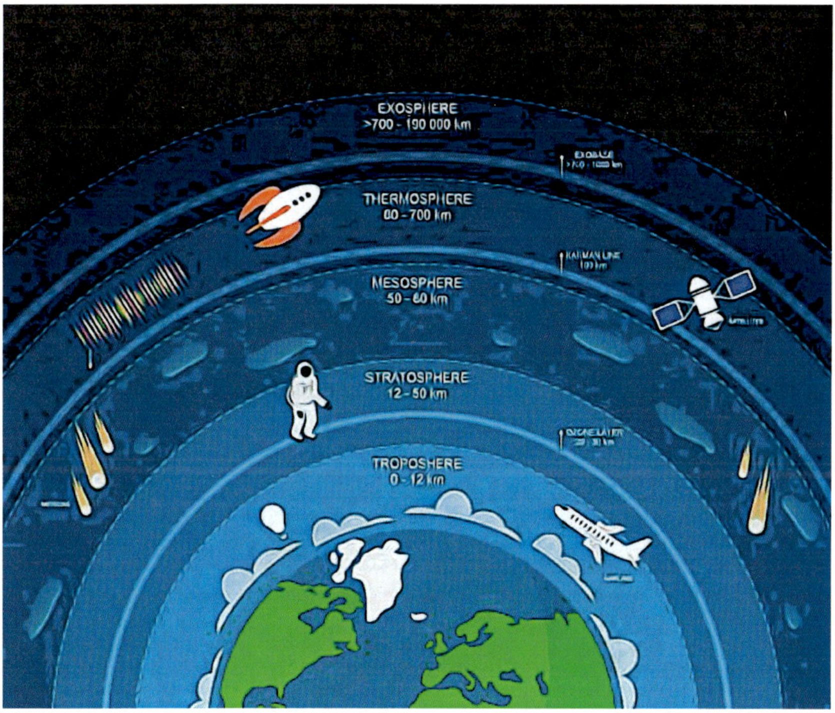

When was the First Planet Discovered?

The first planet to be discovered using a telescope was Uranus, which was recognized as a planet in 1781, based on telescope observations by Sir William Hershel (UK) and others.

When did Americans Start Brushing their Teeth?

Most people started brushing their teeth by the 1930's. In the early 1900's only 7% of American household brushed their teeth or at least had toothpaste in their houses. During World War 1 most of the Army recruits had such poor oral hygiene that the military considered dental disease a national crisis.

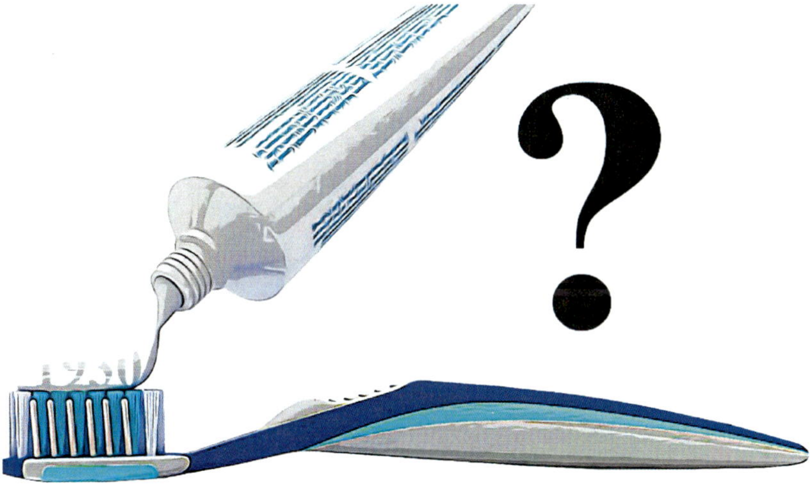

What are Some Odd Names for Individual Cats?

A male cat is referred to as a tom, which you may have heard before, but, a neutered male cat is known as a gib, which, if you knew that one, you should try out for Jeopardy.

What was Martin Luther King Jr.'s Birth Name?

King was born Michael King Jr. on January 15, 1929. In 1934, however, his father, a pastor at Atlanta's Ebenezer Baptist Church, traveled to Germany and became inspired by the Protestant Reformation leader "Martin Luther". As a result, King Sr. changed his own name as well as that of his five-year-old son.

What are the Two Body Parts that Never Stop Growing?

While the rest of our body shrinks as we get older, our **noses, earlobes and ear muscles** keep getting bigger. That is because they are made mostly of cartilage cells, which divide more as we age.

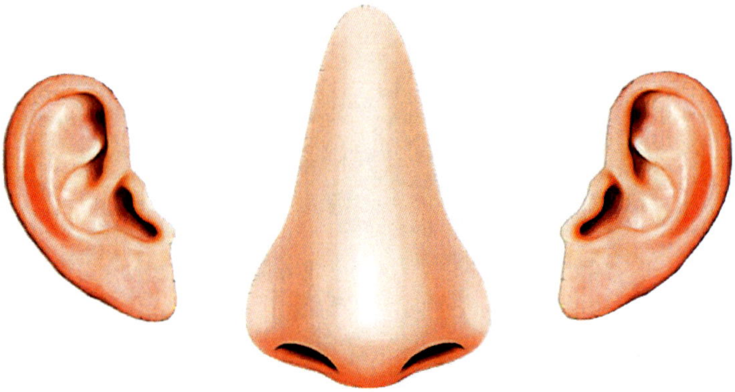

What is a Group of Crows Called?

A group of crows are called a "Murder."

Crows are members of the Corvidae family, which also includes ravens, magpies, and blue jays. Loud, rambunctious, and very intelligent, crows are most often associated with a long history of fear and loathing. They are considered pests by farmers trying to protect their crops and seedlings. Many people fear them simply because of their black feathers, which are often associating them with death. But research demonstrated in a Murder of Crows proves crows are actually very social and caring creatures, and also among the smartest animals on the planet.

Which is the Only Body Part that is Fully Grown from Birth?

The eyeball is the only organism that does not grow from birth. Instead, it is fully grown when you are born.

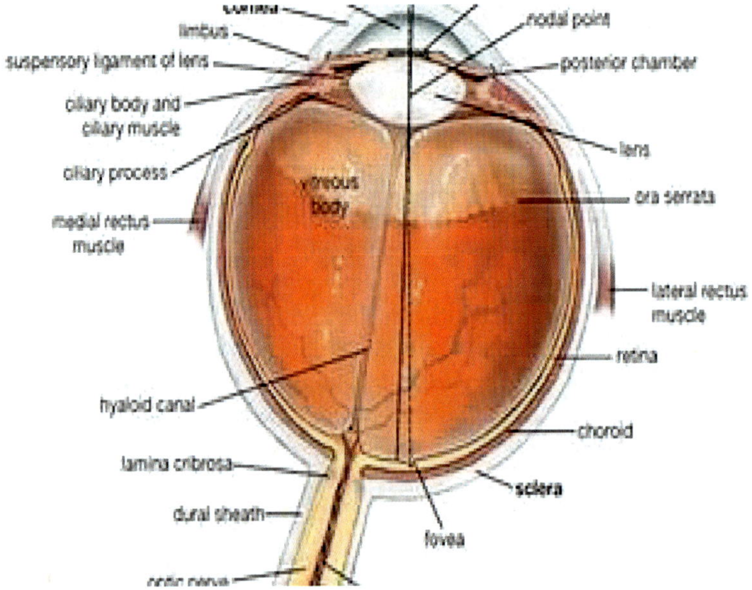

Who are These Inventors? What did They invent?

Lonnie George Johnson is an American inventor, aerospace engineer, and entrepreneur, whose work includes a U.S. Air Force-term of service and a twelve-year stint at NASA, where he worked at the Jet Propulsion Laboratory.

Alexander Miles, who contributed to the elevator industry, was an African American inventor of the late 19th century who transcended racial barriers in the United States. Miles, born in Duluth, Minnesota, designed an elevator that was able to open and close its own doors and the elevator shaft doors.

Marie Van Brittan Brown worked as a nurse and later became a famous inventor. However, she became an iconic figure, not because of her profession as a nurse, but because she invented the audio-video alarm system.

She invented this together with Albert Brown, her husband. This innovation was Marie's solution to her mounting fear of their neighborhood. At the time, it had a remarkably high crime rate, and the police response to emergency calls was incredibly slow.

Marie's home security system laid the foundation for other versions of various security systems that the world uses today in homes, businesses, and some private and public places.

George Washington Carver was an American agricultural scientist and inventor who promoted alternative crops to cotton and methods to prevent soil depletion. He was one of the most prominent black scientists of the early 20th century.

With only an elementary school education, **Garrett Morgan** began his career as a sewing-machine mechanic. He went on to patent several inventions, including an improved sewing machine and traffic signal, a hair-straightening product, and a respiratory device that would later provide the blueprint for WWI gas masks. The inventor died on July 27, 1963, in Cleveland, Ohio.

Every time a person crunches into a potato chip, he or she is enjoying the delicious taste of one of the world's most famous snacks- a treat that might not exist without the contribution of black inventor **George Crum**.

Lloyd P. Ray created the dustpan in 1897. A metal plate with a short wooden handle that allowed sweepers to collect trash without getting their hands dirty, Ray's invention quickly became a household staple.

Gladys West is an American mathematician who is renowned for her pioneering work on the development of the Global Positioning System (GPS), which is widely used for navigation, tracking, and location services. West was born on October 27, 1930, in Sutherland, Virginia, and grew up on a farm with her family.

Madam C.J. Walker became the richest African American women of her time thanks to her work with hair inventions.

Born into poverty, the entrepreneur, philanthropist, and activist invented a comb that would straighten curly hair.

Learning Chart (KWL)

What I Know	What I Want to know	What I Learned

The Fastest Man Alive?

The name Usain Bolt is synonymous with Olympic gold medals, speed, success and world records and for many the Jamaican is regarded as the greatest sprinter of all time. Usain Saint Leo Bolt was born on the 21st of August 1986 in Sherwood Content, a small town in Trelawny, Jamaica. He is a World and Olympic champion sprinter, and many people believe he is the fastest person ever, which has earned him the nickname 'Lightning Bolt'!

Learning Chart (KWL)

What I Know	What I Want to know	What I Learned

What is Figurative Language?

Personification:

Giving Human Characteristics to Something That Is Not Human

Examples:

The trees danced gracefully in the wind.

Kaitlyn heard the last slice of cake calling her name.

The busses are impatient in New York.

The flowers nodded in the wind.

The clouds pushed each other around in the sky.

Onomatopoeia:

Words That Mimic Real Life Sounds

Examples:

The train goes, “choo, choo”.

The car goes “vroom, vroom”.

“Squeak” went the nail on the chalkboard.

“Screeeeeech” went the car as it came to a sudden stop.

“Click” went the car keys.

Hyperbole:

An Exaggeration So Dramatic That No One Would Believe It

Examples:

I'm so hungry I could eat a horse!

I'm so tired I could die!

She's so loud she is waking up the whole world!

I'm dying of boredom!

Your bag weighs 500 pounds!

Alliteration:

When A Set of Words Uses The Same Letters Repeatedly

Examples:

Sally sold seashells by the seashore.

How much wood could a wood chuck chuck if a wood chuck could chuck wood.

Donald the dalmatian dog dug all day.

A big bully beats a baby boy.

But a better butter makes a batter better.

Simile:

A Comparison Using Like or As

Examples:

She is as hungry as a cow.

He eats like a pig.

Angel J. is like a Greek goddess.

Murshid is as strong as the hulk.

Idiom:

A Phrase That Is Not Intended To Be Taken Literally

Examples:

It's raining cats and dogs.

He won by the nose.

If the shoe fits, put it on.

The politician killed two birds with one stone.

Metaphor:

Two Unlike Objects Compared (Without Using Like Or As)

Examples:

Daniel's a pig.

Robin's a pretty bird.

Ahmad's a shadow. Everywhere you go, he follows.

David's the hulk.

Learning Chart (KWL)

What I Know	What I Want to know	What I Learned

What are Reading Strategies?

PREDICT: Determine what you think will happen in the text. Use the title, text, and illustrations to help you.

VISUALIZE: Create mental images of the characters, settings, and events in the text.

QUESTION: Stop and ask yourself questions to see if the text makes sense. Reread the text if you need more information.

CONNECT: Think about what you already know about the text. Find ways to relate the text to yourself. Other text, and the world around you.

INFER: Use clues in the text and your own knowledge to fill in the gaps and draw conclusions.

EVALUATE: Think about the text as a whole and form opinions about what you read.

Learning Chart (KWL)

What I Know	What I Want to know	What I Learned

Getting to Know Poems

Rhyme Scheme:

A rhyme scheme is the pattern of rhymes at the end of each line of a poem or song. It is usually referred to by using letters to indicate which lines rhyme; lines designated with the same letter all rhyme with each other.

Rhyme:

A rhyme is a repetition of similar sounds in the final stressed syllables and any following syllables of two or more words. Most often, this kind of perfect rhyming is consciously used for a musical or aesthetic effect in the final position of lines within poems or songs.

Free Verse:

Free verse is an open form of poetry, which in its modern form arose through the French *vers libre* form. It does not use consistent meter patterns, rhyme, or any musical pattern. It thus tends to follow the rhythm of natural speech.

Couplet:

A couplet is a pair of successive lines of metre in poetry. A couplet usually consists of two successive lines that rhyme and have the same metre. A couplet may be formal or run-on. In a formal couplet, each of the two lines is end-stopped, implying that there is a grammatical pause at the end of a line of verse.

Sonnet:

The sonnet is a type of poem that has been a part of the literary repertoire since the thirteenth century. These literary devices can communicate a sundry of details contained within a single thought, mood, or feeling, typically culminating in the last lines.

Limerick:

A limerick is a form of verse, usually humorous and frequently rude, in five-line, predominantly anapestic trimeter with a strict rhyme scheme of AABBA, in which the first, second- and fifth-line rhyme, while the third and fourth lines are shorter and share a different rhyme.

Narrative Poetry:

Narrative poetry is a form of poetry that tells a story, often using the voices of both a narrator and characters; the entire story is usually written in metered verse. Narrative poems do not need to rhyme. The poems that make up this genre may be short or long, and the story it relates to may be complex.

Alliteration:

Alliteration is the conspicuous repetition of initial consonant sounds of nearby words in a phrase, often used as a literary device.

Verse:

A verse is formally a single metrical line in a poetic composition. However, verse has come to represent any grouping of lines in a poetic composition, with groupings traditionally having been referred to as stanzas. Verse in the uncountable sense refers to poetry in contrast to prose.

Learning Chart (KWL)

What I Know	What I Want to know	What I Learned

Words from My Heart in the "Form of Poems"

MURSHID

My Son,

My First Born

18 years ago, we meet,

your tiny body, small hands.

I look at you today son and see a man!!

Through the years our relationship has been

Sunny, Rainy, Cold, and a little bit Snowy,

but no matter the weather we've kept our love for

each other together.

We haven't always seen eye to eye but to see you

today, I realize my goal has been accomplished.

It hasn't been easy but, thank you it hasn't been so

hard. I've had help my son, you didn't know, I

asked GOD from day one to keep you for me and to guide you as only he could do.

We've come a long way together you and I, you've seen me through my childhood, Fifteen I was when you were born.
You've kept me up a many of nights, your cries
I 've heard your tears, I've wiped.

Now it's time for us to part, oh my son I'll miss you with all my heart. Let your life be your elevator, always going up and in your travels keep GOD near for he's the one to call on when MAMA can't hear.

Son, I know this might sound strange, and you might even laugh but: I don't have another word to say except I'm proud of you.

Home Forever Home

MAMA

MONICA RENA

GOD is always Good,

GOD is always Giving, and to us, He gives you

A Life inside your mother not yet born,
But as the days turn to night, a summer takes flight
Winter will come and so will you, my precious little one

Monica Rena, I'm told your name will be
But no matter the name, you're a part of me.
Your father's mother I am.

For I am patiently waiting to meet you,
Patiently waiting to hold you,
For I am patiently waiting to love you

GOD is a good GOD,
His gifts are grand, His love is true

For it is He, that has given us you!!!!

- grandma
DELISSA RENA
November 11, 1998 (5:00PM)

My Sweet Little Angel

Well today is the day you were born

15 years ago

I should be happy, but instead I'm sad

I feel lonely for the day my Angel *will go away*

What will I do without my Angel *to*

Hug me when I'm feeling low?

What will I do without my Angel *to*

Rub me and make the pain go away?

Oh my GOD what will I do

my son has gone, out on his own

my angel is soon to follow

I should be happy

I should be proud and both I am

But it's my heart

I can't tell my heart to stop feeling lonely

I can't tell my heart to stop feeling sad

I can't even tell these old eyes of mine to stop

crying

my sweet little Angel

I love you with all my heart and soul I hope you know

the day will come you'll have you an angel

then you will know the things my heart is saying

that words could never say

My sweet little Angel

CHANGE, ARE YOU READY?

If you are not ready for change

You need to get ready

As the egg that makes a baby does not remain

an egg

An the baby that comes from the egg does not remain

a baby

focus in this and know that life in itself is

CHANGE

If you feel your life is all-right

an you need nothing to change

check your pulse my friend: for you might be dead

Delissa R. Frankin

Yes I'm Ready

Learning Chart (KWL)

What I Know	What I Want to know	What I Learned

What is?

Author's Purpose! 

Author's purpose is the main reasons he or she has for writing, (persuade, inform, entertain).

What is the difference? Connotation vs Denotation

A connotation is the feeling a word invokes. But take note! A denotation is what the word literally says. If these words were on a trip, connotation would be the baggage, and denotation would be the traveler.

A connotation is the baggage a word or idea drags around. The word "Baggage" often has a negative connotation. If you say someone has baggage. Unless they're at the airport, you mean the person is lugging around some drama.

What is? A thesis statement in an essay

A thesis statement is **the main idea of an essay.** It consists of the topic of the essay and the writer's claim about the topic of the essay and the writer's claim about the topic that will be proven throughout the essay. The thesis usually appears at the end of the introduction, often as the last sentence, and lets the reader know what to expect.

Example: Peanut butter and jelly sandwiches are the best type of sandwich because they are versatile, easy to make, and taste good.

In this persuasive thesis statement, you see that I state my opinion (the best type of sandwich), which means I have chosen a stance.

What is? Making an Inference

Making an inference simply means to use what you already know to make a guess about something that you do not know because things aren't always told to you directly.

Learning Chart (KWL)

What I Know	What I Want to know	What I Learned

Did You Know?

Tooth Enamel is the hardest Substance in the human body!

Move over, Bones! The outer layer protecting our teeth is the hardest thing in our bodies. The next layer down, dentin, is also stronger than bone. The trade-off is that teeth have a very limited ability to heal themselves, unlike bones. You can't put a cast on a cavity, after all.

Snails have thousands of teeth!

Each unassuming snail hides a microscopic secret: between 1,000 and 12,000 tiny teeth protruding from its tongue. They use these teeth to break down parts of their food, and since the teeth are not especially durable, they need to be replaced pretty frequently. This tooth-covered tongue is called a radula, and it's not exclusive to snails. Slugs have them, too, along with some squids.

Race is what you see (physical traits):

- Skin color
- Hair texture
- Facial features
- Eye color

"A person does not choose race. It is assigned by society (socially constructed) based upon these physical features," In other words, your race is determined by how you look. The U.S. Census Bureau outlines five categories of race:

1. Black or African American
2. American Indian or Alaska Native
3. Asian
4. Native Hawaiian or other Pacific Islander
5. White

All five races are determined by how you look. Race is a social construct, which means it exists because humans agree it exists. However, just because it's a construct doesn't mean it doesn't have real implications. For example, racism exists due to classifying and treating people differently based on their grouping.

Ethnicity is a fluid and self-determined based on the social and cultural groups you identify with through learned aspects of your life. While Merriam-Webster Dictionary defines 'ethnic' as relating to large people groups with the following commonalities:

- Racial
- National
- Tribal
- Religious
- Linguistic
- Cultural background

Dr. Baxley specifically explained it simply as, "Ethnicity is the cultural practices, customs, and traditions that are learned and shared by a group of people—including language, nationality, heritage, religion, dress, and customs."

Dolphins Have Three Stomachs

While dolphins are conning hunters, still, they forgo any semblance of dining manners by wolfing down their food without chewing. Thankfully, they have a three-chambered stomach designed to handle the rapid intake. Their easily stretchable forestomach is primarily used to store the food. However, the main stomach is where the majority of digestion take place, and the pyloric chamber completes digestion and regulates passage into the small intestine.

Oprah Winfrey, the "Queen of All Media"

Her real first name is "Orpah." According to her birth certificate, we've been saying the media queen's first name incorrectly all this time. Born in January 1954 in Kosciusko, Mississippi, the future TV host was named "Orpah," after a woman in the Bible's Book of Ruth. However, the unusual name immediately caused confusion among family, who adopted a slightly different version of the moniker for their newest member. As she later explained during one early audition tape, "No one know how to spell in my home, and that's why it ended up being Oprah."

Learning Chart (KWL)

What I Know	What I Want to know	What I Learned

Do You Know?

Do you know? What you want to be when you grow up?

When I grow up, I want to be

______________________.

Do you know? If you want to go to college after you graduate from high school?

After I graduate from high school I want to

_______________________.

Do you know? Where will you live one year after you finish high school?

One year after I finish high school I plan to live

______________________.

Do you know? Where will you live five years after you finish high school?

Five years after I finish High school I plan to live

______________________.

Do you know? What makes you happy?

One thing that makes me happy is

_________________.

Learning Chart (KWL)

What I Know	What I Want to know	What I Learned

HMM

If you have good thoughts, they will shine out of your face like sunbeams and you will always look lovely!
Roald Dahl

dreams

you're

AMAZING

just the way

YOU ARE

ALWAYS
BELIEVE THAT
SOMETHING
Wonderful
IS ABOUT
TO HAPPEN

Never Never Never Never
Never Never Never Never
EVER EVER EVER
EVER EVER EVER
EVER EVER
GIVE UP
Stay Strong

ALWAYS
SEEK
KNOWLEDGE

FRESH
DAY,
FRESH
Start

“

Nothing is impossible,
the word itself says
‘I’m possible.’

AUDREY HEPBURN

Southern Living

TODAY A READER tomorrow a LEADER